INTRODUCTION

This booklet has been produced by the
tion to supply the need for a simple a
stone walling.

It is designed to help those who wish
stone walling into effect for themselv
dens or on their farms; to help the
oldest craft; or for those who are int
walled landscapes of Britain. It is al ...panion for those who wish to attend one of the many short courses now run by branches of the Dry Stone Walling Association, which are immensely valuable to people with no previous practical experience.

The Dry Stone Walling Association

The Dry Stone Walling Association was formed in 1968 to counter the then almost terminal decline in the craft. As a result of its efforts on behalf of professional and amateur members alike, there has been a substantial upturn in the fortunes of dry stone walling. Members receive a magazine, the *Waller and Dyker*, three times a year which contains general interest and specialist articles, accounts of branch activities, details of short courses and contact addresses. The Dry Stone Walling Association operates a Craftsman Certification Scheme at three levels. A holder of a Master Craftsman Certificate will turn out professional work under all circumstances. In addition, the Association publishes a *Register of Professional Wallers and Sources of Stone* which is circulated widely to those who employ wallers such as Regional and District Authorities, Civil Engineers, etc.

Members are always welcomed to support this ancient craft and subscriptions are modest. If you would like further information about any aspect of the Association's work please write to:

The Secretary, Dry Stone Walling Association of Great Britain,
YFC Centre, National Agricultural Centre,
Kenilworth, Warwickshire CV8 2LG

WHY HAVE DRY STONE WALLS?

Dry stone walling involves building in stone without the use of mortar or other binding material, and the technique is extensively used in large areas of Britain and throughout the world. This form of construction has many advantages. The traditional material used was essentially free and helped in clearing stone waste from fields. It uses natural local materials, is fireproof and it can be used where other materials can not, for instance across rocky ground where fence posts cannot be driven.

A dry stone wall will settle naturally instead of cracking apart, and it is not normally affected by water. Walls provide shelter and shade to livestock, and give a longer crop growing season on the land adjacent to them. They provide a splendid habitat to many species of small birds and animals, insects and flora; and because the materials used are natural, the walls blend harmoniously into the landscape. In the garden, dry stone walls form a splendid backdrop for roses, climbing and alpine plants.

No other form of enclosure comes close to providing such a formidable list of advantages!

SOURCES OF STONE

The ease and expense of finding suitable stone depends largely upon where you live. If you live in a walling area, which in practice is most of Britain apart from the south east, the best place to obtain stone is from a derelict field wall for the stone will have been sorted and worked into suitable types generations ago and will be attractively weathered. A word with local landowners should produce results, but be prepared to pay them something and to come to an arrangement over haulage.

Disused quarries may have heaps of waste stone that is usable and which the owners may allow you to take. In some coastal areas the shore provides excellent building stone. Always check with landowners and with local authorities before taking any stone - some areas are "protected" and you could be breaking the law to remove the stone!

Dry stone walling requires a great deal of stone - a metre of wall approximately one and a quarter metres high takes about a tonne of stone. Buying stone from a commercial quarry can be very expensive, but where a particular stone type or colour is required it may be your only option. The *Register of Professional Wallers and Sources of Stone* published by the Association provides some information. Also, the Stone Federation - Great Britain, 82 New Cavendish Street, London W1M 8AD (071 580 5588) can provide a list of working quarries operated by their members.

TOOLS AND EQUIPMENT

The tools needed are simple and inexpensive, but the following are more or less essential:-

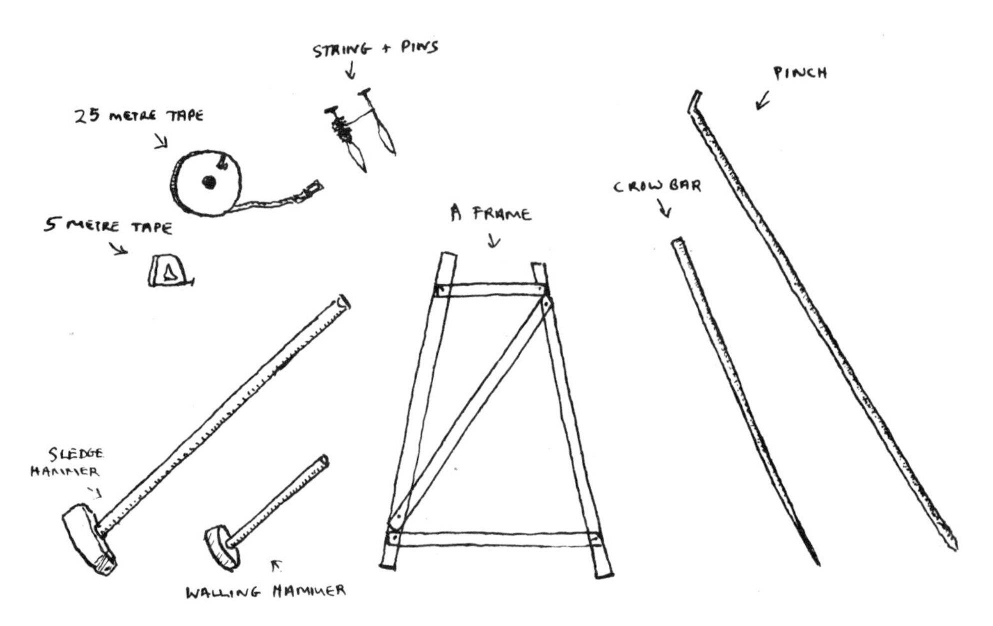

Diagram 1 - Tools and Equipment

1. *A measuring rule*

2. *Building Lines:* made of tough string or nylon. Two are needed, at least 10 metres long with flat metal pins which can fix into the wall.

3. *Walling Frames:* these are wooden frames which have the same "cross section" dimensions as the wall that is to be built. Any wood may be used, provided the outer edges are straight, about 75mm x 75mm is ideal. It is important to make them symmetrical and true, and to make sure they are completely vertical when in use. Two will be needed to guide the strings, which are raised up the frames as the wall is built.

4. *Pick and/or Crowbar:* these are used for moving large or half-buried stones.

5. *Hammer(s):* a stone hammer weighing 1.5 to 2 kg is required for some shaping of stone and for breaking waste stone for filling. A larger 6kg hammer is also useful for the larger stones if available.

SAFETY

Take great care when hitting stones with a hammer - small pieces of stone can break off at high speed and are particularly dangerous to the eyes. Turn the head or blink as the hammer hits the stone, or best of all, wear inexpensive, plastic safety spectacles or goggles.

With heavy stones, be careful not to trap your fingers or toes, and always lift by bending the legs and keeping the back straight. When removing heavy stones off an old wall, stand tight to the wall with your hips against it. If in doubt, always ask a neighbour to help with the lifting. Most professional wallers wear gloves to protect them, especially in winter when a combination of rough stone and wet conditions can rapidly play havoc with the skin. Industrial rubber gloves are the most durable, and cheap.

Work boots or Wellington boots with steel toe caps also make sense for although serious injury to the feet is rare, even a small stone falling on to a toe joint can be extremely painful!

DIMENSIONS

The dimensions used in this booklet are for a wall 1.40 m high which is about right for stock proofing a field. However, a wall can be built more or less any convenient height, and a rough guide is that the width of the wall under the "top" or copestone should be half that of the base. If in doubt, make the wall wider rather than narrower.

Note: the stones placed along the top of the wall will be referred to as copestones in the remainder of this booklet.

BUILDING THE WALL

There are several different styles in which a wall may be built - the finished look will depend on which stone is available to you. The wall described in this publication relates to areas where there is a high proportion of smaller stones. The use of larger stones for single walling is described in books for the more advanced waller.

Sorting the stone

Much time and frustration can be saved if a certain amount of sorting of stone is done before building begins. Obviously, if one is dismantling an existing wall the work has already been done, and it is a relatively simple matter to place the stone removed in a sensible way adjacent to the wall that is to be rebuilt. This will be described shortly.

There are five basic types of stone described below:

1. **Copestones** These are placed at the top of the wall, tightly together. They should be regular in shape, at least 20 cm high, and you will need 7 or 8 for each metre length of wall, more in some districts. Be careful not to use them elsewhere in the wall.
2. **Building stones** These stones (often called "the double") are used in the face of the wall. Larger stones should be used in the bottom layers, where possible, smaller ones towards the top.
3. **Throughstones** Stones which are placed about half-way up the wall and go right through, binding the sides together. One will be required for each metre of wall. Their length makes them special stones, so do not break them for other purposes.
4. **Foundation stones** These are heavy flattish stones, used right at the base of the wall and are especially important because the whole weight of the wall rests on them.
5. **Filling** Also known as "hearting", this is small waste stone used to pack the middle of the wall between the two faces.

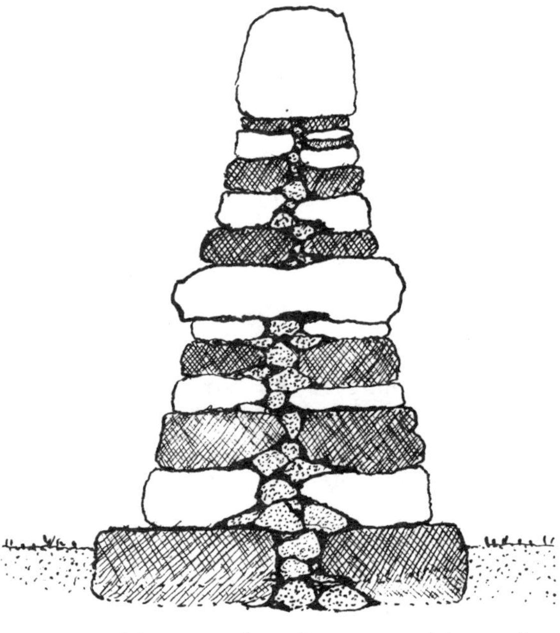

Diagram 2 - Cross-section of a finished wall

When dismantling an old wall, also referred to as "stripping out", much can be learned by an intelligent appraisal of how it was originally built, and the way the different stone was used. The practice of stripping out an old wall is carried out in the following manner.

First, remove the copestones, and place in a line some two metres back from the wall.

Then, place the building or double stones as close to the wall as will allow room to work - aim for some 15 cm clear between the foundation-line and the edge of the pile of stone. Make sure there is an equal amount of stone on either side of the wall.

Gather together the small stones for packing and hearting and place them beside the building stones on both sides of the wall.

Place the throughstones on the opposite side of the wall to the copestones about the same distance away.

Unless built on rock, all walls sink with time. In many cases it may not be necessary to remove the old foundations from below ground level. If it is firm and level, leave it but err on the side of caution and in particular, prise out any that have slid or tilted away from the line. The heavy stones used in the foundation should be left immediately adjacent on either side of the wall.

STAGE ONE – THE FOUNDATIONS

Mark out the line of the wall with the strings (which should be attached to simple pegs) 85 cm apart. If on softer ground, dig a shallow trench so that loose earth or grass is removed. On stony ground this is not necessary so long as the site is level and without such obstructions as loose stones or tree roots. Now, reposition the strings to 80 cm apart, ready for the foundation stones.

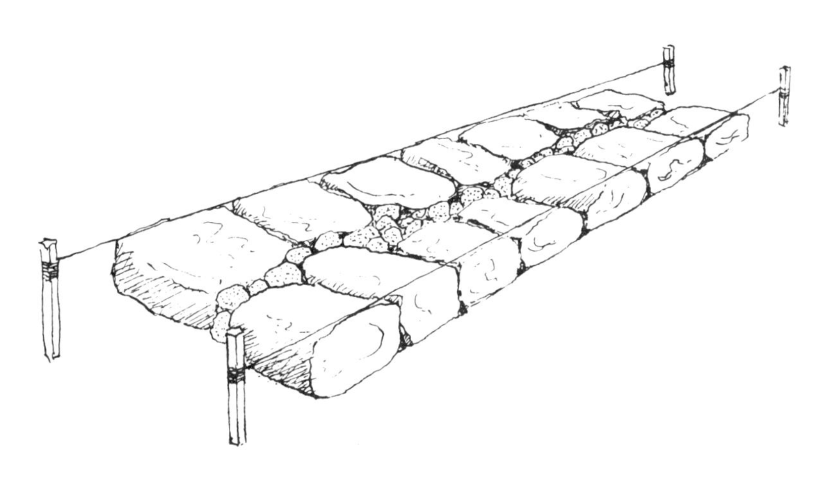

Diagram 3 - Foundations

The foundation layer is wider than the base of the rest of the wall, as this increased surface area at the bottom makes for a more stable structure and the finished wall will also settle more slowly. The foundation stones should be flattish and reasonably large. Small stones will be pressed into the ground and a mixture of large and small stones will give uneven settlement. Try also to match the stones for height to give as level a surface as possible on which to build the rest of the wall. Ensure they are lying securely and cannot be moved when walked on.

The entire foundation of the length you have laid string along should be finished before going further. This also applies to building the rest of the wall - complete each course before proceeding. Some or all of the above stages may not be necessary where a wall is being rebuilt.

Key Points of Building

It is important to stress two rules at this stage. Firstly, *always* make sure that each stone touches the one beside it. Secondly, *always* fill up the centre of the wall carefully as you build. The whole strength of the wall depends on these two rules being carefully observed. There are numerous examples of walls collapsing after a few years due to lack of care in these matters.

When building the wall, you should also be aware of several further points. When laying the building or face stones, try to keep the stones level. This is done by lifting where necessary the rear part of the stone and slipping a wedge underneath. Make sure you do not leave holes or spaces under any of your stones. Place your stones as close as you can to the strings <u>without</u> actually touching them. If the strings are touched they will move outwards, and even a small movement will alter the shape and strength of the wall. The filling (hearting) must not be just dropped or thrown into the centre but carefully packed so that they will not move easily. A good test of whether you are building the wall correctly is to stand on it at any time - the stones you have laid should not move under such pressure. If you follow the above rules - and they are not difficult with a little practise - then your wall will be strong and last for many years.

STAGE TWO
BUILDING THE LOWER COURSE

If you are not starting to build at a wallhead (where the wall begins, and which is self-supporting, see page....) set up two wall frames about 5-10 metres apart. Support them with a simple post of wood or metal. If these are not readily available, then with care the frames can be supported with a few carefully positioned large stones. Whatever you do, make sure the frames are straight, level and vertical. Tie the strings to the frames so that they are about 15 cm above the foundation layer, and make sure they are tight and do not sag.

Observing simple rules will ensure a strong, good-looking wall.

1. Most importantly, *always* cover the joints formed by the stones below otherwise lines of weakness will develop and bulging may occur. The golden rule is: *one stone on two, and two stones on one.*

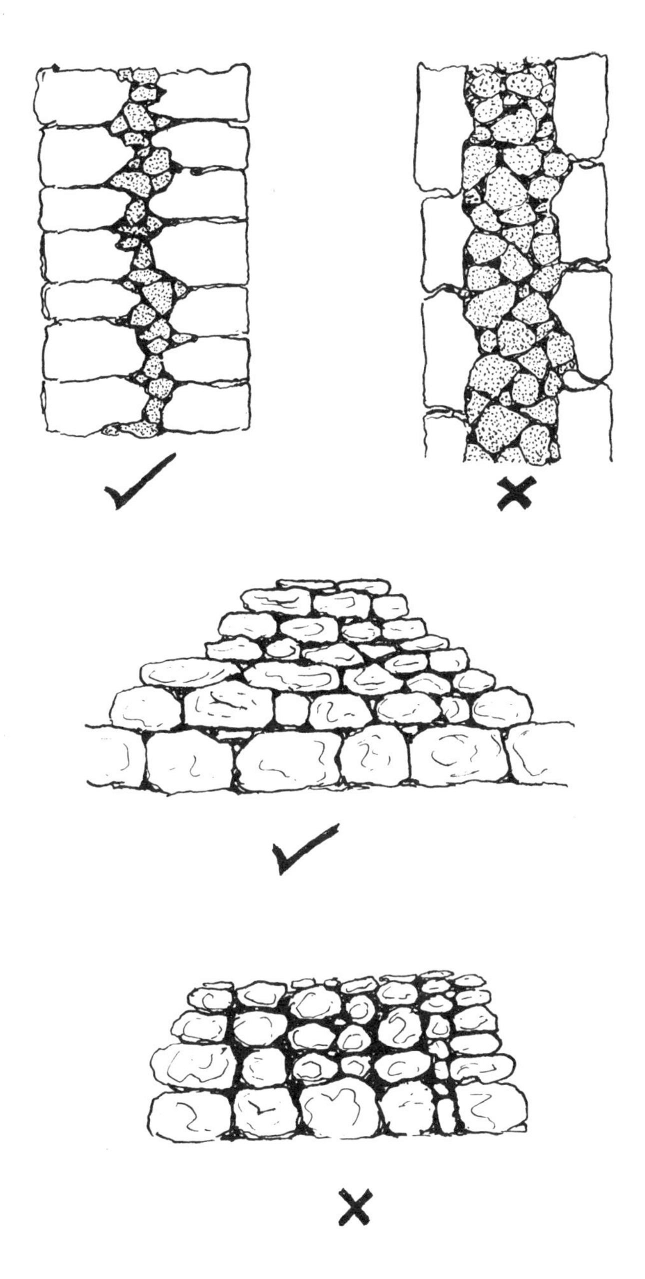

Diagram 4 - Rights and wrongs

2. Lay the stones so that they lie with the longer side running into the wall, not along the face. This gives a much stronger wall.

3. Build up course by course so that when you have reached (or gone above) the strings at 15 cm, raise the strings another 15 cm and so on.

4. Make sure that each stone is firmly in place before moving on to the next. Stones do not tighten up as further stones are laid on top. Do try and resist the temptation to leave gaps because the stone in your hand fits better a little further along the wall. Unless you are very experienced, it is far easier to lay the stones touching one after the other than to have to find a specific stone to fill a specific gap.

5. As far as possible, use the larger stones at the bottom of the wall. If used too near the top, there is the danger that there will not be sufficient space left on the opposite side of the wall.

6. Fill the wall as the work proceeds - do not leave it as a separate job.

7. Build up both sides of the wall at the same time, keeping each side roughly the same height.

Diagram 5a - First lift, levelled and ready for throughstones.

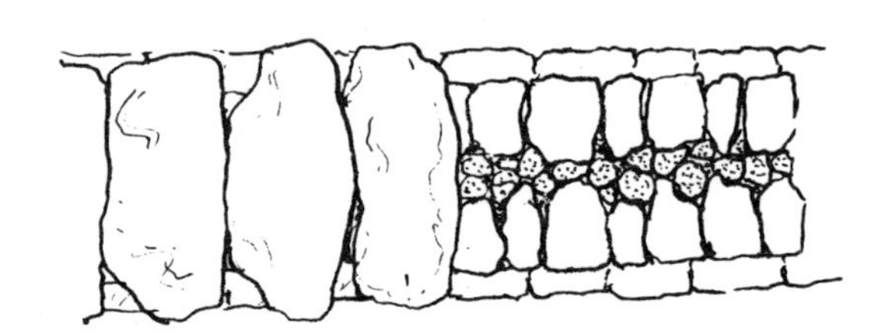

Diagram 5b - Throughbands may be laid closer together if abundant

STAGE THREE – THE THROUGHSTONES

You will remember that the throughstones are placed about halfway up the wall, between the foundation course and the bottom of the copestones. Assuming the height of the wall to be 1.4 m, and allowing about 20 cm for the copestones, this leaves about 1.15 m for the courses. therefore the throughstones should be placed at about 0.55 m.

The throughstones should go through the wall, but should not project more than 5 cm on either side. If they stick out further, animals may use them to rub against and hence loosen the wall - or people may hurt themselves when passing. Having said this however, they are better too long than too short.

Having attached your strings to the wall frames at a height of 0.55 m, level the wall at this stage. Use flattish stones if necessary, as shown in the diagram. This done, place the throughstones carefully on the wall at not more than 1m intervals. You may, however, place them closer together if you have many such stones, and take care when lifting them as they will be heavy. Do not bang them down on the wall, or slide them across: both actions will loosen the stones below. If necessary, wedge small stones underneath to prevent them rocking.

STAGE FOUR – SECOND LIFT

The strings are now raised again, and the courses are continued for a further 0.55 m or so. As the wall goes up it becomes considerably narrower, which is why the larger stones should be used in the lower courses. It is even more important to pack these smaller stones very carefully as they will move more easily unless completely firm. When the strings are raised to their final height, level off accurately to give a pleasing "clean cut" look.

STAGE FIVE – THE COPESTONES

The wall will now be about 35 cm wide at the top of these courses, and will now be ready for the copestones. These will not only bind the wall together at the top in the same way as the throughstones do in the middle, but will resist damage if the wall is crossed by man or animal. It also makes the finished wall look attractive and well built.

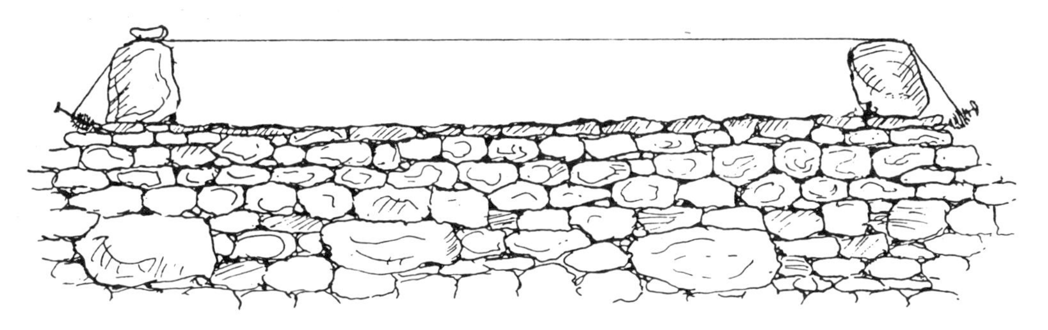

Diagram 6 - Strings set for copes

Place a copestone at each end of the stretch you are building and make sure they are secure. Then run one of your strings tightly between them as your guide for finished height. The angle at which these copestones are laid should perhaps reflect the way they are used in your area. Some parts of Britain place copestones vertically, in others they are laid at an angle of approximately 45 deg. Whichever you decide, lay one copestone after the other, so that each one is almost, but not quite, touching the string. If any stone does touch the string, it will force the string up and the wall will become uneven. Some of your copestones will be too low, in which case place a flattish stone under one or both sides of the copestone to bring it up to the required height.

Having topped-off the wall, the copestones are now "pinned". This involves filling the larger spaces with wedge shaped stones to prevent any rocking or twisting, and locks the copestones so that they are completely immovable. This is done by taking "V" shaped slices of stone and driving them into the spaces between the copestones with your hammer until they are firmly wedged. If this is done properly, then you will be able to walk along the top of the wall without any movement of the stones below.

Finally, when the job is complete tidy up by gathering together any unused stone into small heaps which can be later removed.

--- oOo ---

BUILDING IN DIFFERENT STYLES

Some areas of the country, for instance the limestone areas of Yorkshire or the Cotswolds, are built entirely of building, or double, stones. Others, such as parts of the West Country, the North East of England and Scotland have walls that are partly or wholly constructed of larger stones and boulders.

This publication does not cover the more complicated procedures involved in some regional styles: singling, Galloway-dyking, etc. These will be described in another booklet published by the Dry Stone Walling Association.

BUILDING ON STEEP SLOPES

Gentle slopes can be built on without the employment of any special techniques. With slopes up to an angle of around 15 degrees, it is customary to lay the "double" parallel to the slope, so that the stone follows the undulations of the ground.

At angles steeper than this, the foundations must be formed into a series of little steps. Starting at the bottom of the slope, a horizontal "bed" is formed for the first stone. The earth adjacent to this stone - above it - is levelled for a short distance, so that the next stone is laid half on the one below, and half on the compacted earth. It is like forming a series of small steps, and it is important to lay all the stone horizontally as in the diagram.

When the double has reached perhaps halfway up, place on a large stone, and build away from that up the slope. If large stones are not available, build the double to the top, and step it in a similar way to the foundations.

Diagram 7 - Steep slope and "Galloway" wall

When using large stones, ensure that they are placed either vertically, or leaning *into* the slope.

When placing the copestones, you must start at the top of the length you are working on, and proceed one at a time down the slope. This ensures that each copestone is self-supporting. Should one of these be knocked off later, it will not cause all those above it to fall. It is all too easy, should the mistake be made of working up the slope, to lean each copestone against its neighbour below. This is incorrect practice.

One last point. Take great care to place the stones you will be using securely on the ground. It is both annoying and dangerous to people below, for them to roll off down the hill!

CHEEKS / WALLHEADS

Unless one is building the wall between two structures, say a house and a garage, it will be necessary to begin at a wallhead (also known as a cheek). Wallheads will also be necessary where gates are located.

Diagram 8 - Doubled Wallhead

The diagram illustrates a "doubled" wallhead. The frame is positioned with a stake, and checked for level. As in the diagram, carefully chosen stones are built up layer by layer, so that stones stretch first of all *across* the wall, then alternate with stones stretching back *into* the wall. Take care to follow the required taper, and as the stones required may be in short supply, always put aside a suitable number and refrain from breaking good wallhead stones for use elsewhere in the wall.

It is good practice to build up the wallhead as the rest of the wall proceeds, rather than construct it all at once.

Since wallheads are so important, if large stones are available in your area, then hunt out the largest stones which can be manoeuvred. These should have as regular a shape as possible - less common than might be supposed. Perhaps only four or five may be needed to provide the strongest form of wallhead possible. It is vital to ensure that every other stone ties back into the wall however, so the wallhead will not split away from the rest of the wall in time.

CORNERS

Corners are used where walls turn sharply, usually at right angles, and are so built as to ensure that two walls tie firmly into each other.

Corners are a little more tricky to erect than a wallhead, as there are extra "angles" to think of in order to maintain the batter and taper of each wall. However, stones are laid alternately stretching first into one wall, then the other as shown in the diagram, both on the external and internal angles. Yet again, the importance of laying the right stone aside for the job cannot be over emphasised.

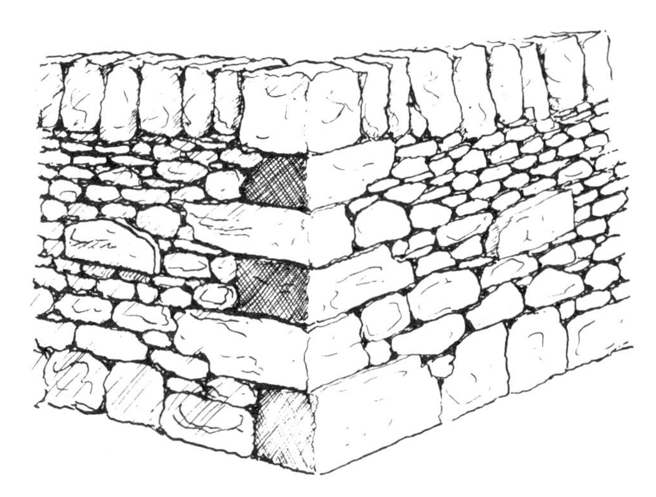

Diagram 9 - Right angled corner

GAPS

Gaps are perhaps the best place to start gaining experience in dry stone work for the beginner, for the stone should be available, the line of the wall is established, and the existing wall will provide plenty of clues as to how to proceed.

If this is used as your starting point, choose a wall that is, apart from the gap, otherwise sound, for it is next to impossible to build against something which is unstable.

Carefully remove the tumbled stone, and place beside the wall. Often, the job will appear daunting, especially where stock have moved freely through the gap for a number of years and there may be earth and weeds mixed in with the stone. However, persevere and do not be discouraged. Once you have worked your way into the gap, the old foundations and maybe some of the lower courses, will appear and may well be sound and in line. Do check, though, why the wall fell in the first place. It may be that the foundations stones have slipped, in which case they must of course be removed and laid again.

The other important point to remember when preparing the gap for rebuilding is that you must also dismantle the wall on either side until you are working against a sound structure. This often means that the gap to be built is much wider than at first appears.

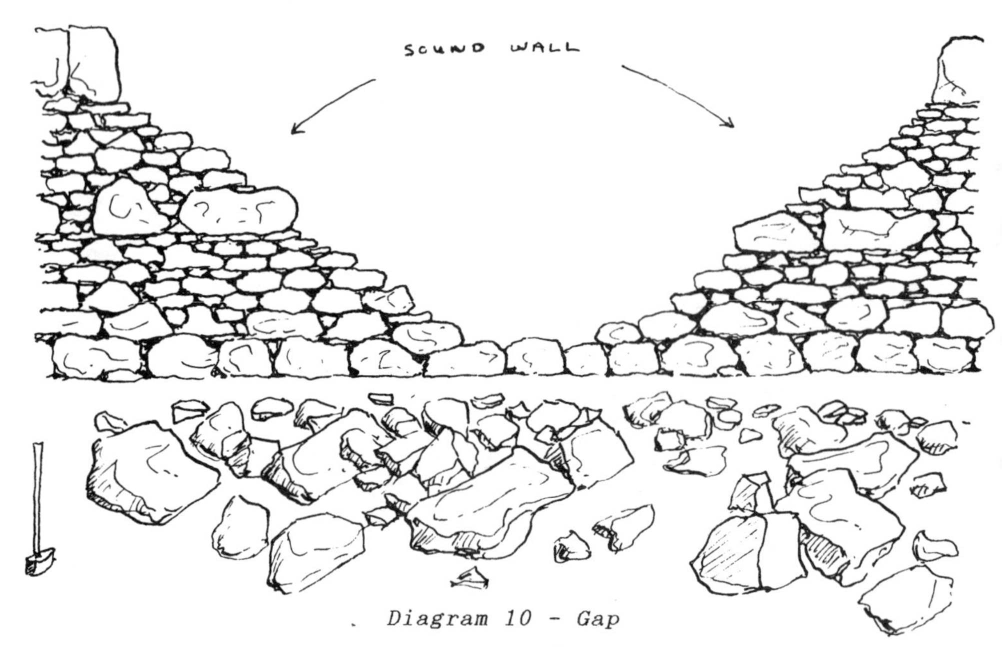

Diagram 10 - Gap

Once all is ready for building, attach the strings into the sound wall on either side and work following the guidelines already described. Ensure that the joints are broken not only in your own work, but also where you join into the old wall. Finally, assume you will need extra material - it is often amazing how stone seems to vanish from an old wall as stones fall out over the years and become buried.

SUMMARY

Dry stone walling can be a most satisfying and creative hobby, and success will be ensured by remembering the following:

** Ensure you have enough stone.

** Take care to sort the stone well and lay out to keep unnecessary running about to a minimum.

** Work slowly and with care, for it is the finished article that matters, not how fast it is built.

** Practice is everything in dry stone walling - expect to be baffled by some of the shapes that confront you. If you are, go and have a cup of tea - the odds are high that the stones will fit on your return!

ADVICE

The **Dry Stone Walling Association of Great Britain** will always be pleased to try and help with advice, answers to technical problems, contacts and information on suppliers. Please note that a stamped, addressed envelope makes life much easier for the secretariat!

FURTHER READING

Dry Stone Walling by Col. F. Rainsford Hannay. This is not a technical manual, but a survey of many aspects of the craft in Britain and overseas illustrated with first-class photographs. There is some basic information on construction techniques. Originally published by Faber and reprinted by the Stewartry of Kirkcudbrightshire Drystane Dyking Committee. Available from the Dry Stone Walling Association.

Dry Stone Walling produced by the British Trust for Conservation Volunteers is a very comprehensive manual which includes historical, geological and regional information. It can be obtained from BTCV or from the Dry Stone Walling Association.

Dry Stone Walls by Lawrence Garner, published by Shire Publications. An interesting book briefly covering the history of the craft and including some guidelines for beginners. Available from the Dry Stone Walling Association.

Building Special Features in Dry Stone by Richard Tufnell and published by the Dry Stone Walling Association. An excellent booklet with construction information for many features - especially those used in gardens. Available from the Dry Stone Walling Association.

Details of currently available titles and mail order costs can be obtained from the Association.

GLOSSARY

"A" FRAME: is a wooden or metal frame used as a guide when building.

BATTER: This is the inward taper of the wall from base to top.

CONSUMPTION DYKE: Wall built with stone to clear the land and which is especially wide. Also called "clearance wall" and "accretion wall".

COPE STONES: the top stones, the stones along the top of the wall to give weight and protection. Also called "cams", "tops", "toppers".

COURSE: Horizontal layer of stones placed in a wall.

COVERBAND: Large flat stones placed across width at top of wall in some areas to form base for the copestones.

DOUBLING OR DOUBLE DYKING: term used for a dry stone wall built with two faces of stones, packed with hearting between.

FOUNDATION: the first layer of large stones in the base of the wall, also called "footing" or "found".

GALLOWAY DYKE: wall or dyke with lower third "doubled", upper two thirds in single walling.

GAP OR GAPPING: a breach in a dry stone wall. Gapping is the repair of same and the "gapper" is the waller or dyker who carries out the repair.

HEARTING: the small stones used as filling or packing in a double wall.

PINNINGS / PINS: small, usually tapering stones used to wedge building stones firmly in place.

RETAINING WALL: dry stone wall built into the cut face of a bank to prevent the soil from moving down the slope.

SINGLE DYKE: wall built with single stones going the width of the wall.

THROUGHSTONES: heavy, large stones placed at regular intervals along the wall to·tie the two sides together.

TRACE WALLING: placing of stones with their length along face of wall rather than placing into the wall for strength.

WALLHEAD: Vertical end to a length of wall. Also called "cheek-end".